My 350 V8 Corvair

A unique muscle car built by,

Ronald S. Craig

ABOUT THE AUTHOR

Ronald (Ron) Craig has spent most of his life in Southern California.

Following his service in the United States Air Force during the Vietnam War, he earned a Business Degree from California State University, San Bernardino while starting a family. He currently resides in Elverta, California, close to four of his six grandchildren that his two daughters have blessed him with.

Ron is currently semi-retired, does engineering cad design work, tutors through Reading Partners Organization at an elementary school. He has also published seven novels, listed in back.

Ron encourages readers to contact him through his author's website,

www.roncraigbooks.com

Chapter 1

Where did the idea of building a muscle car begin? I guess it's always been something that's haunted me. When I was fifteen my sister gave me her old Oldsmobile which I stripped down to the chassis and working three weeks in that summer in my Uncle Paul's auto shop, earned me a 354 Desoto hemi engine and transmission. It was a time when VW, bug bodies were being replaced with fiberglass open bodies to make street dune buggies, so I was able to get the old VW bug body for $25.00.

Shortening the chassis to put the wheels in the right place all the pieces came together. Thank God, I never got it running because chances are I wouldn't still be here to tell you about my Corvair project. There was another prelude to this story and I can confirm a Buick V-6, using a Ford Pinto front chassis will greatly enhance the performance of a MG Midget.

Well, back on track. In March 2009 the muscle car haunting hit me again and the only thing I was sure of was that a modified 350 Chevy engine was going into the mix. My initial thought was to complete the V8 VW bug thing. I did some looking around on Ebay, more focused in a price range than models, keeping options open and came across some Corvairs. I had a 1965 two door Corvair in high school and always loved the body style.

Corvair, you ask? Okay a short history for those of you born in the 80's and later. Chevrolet introduced the Corvair in 1960 to compete with the wave of interest in the VW Beetle, a four cylinder, air cooled, rear engine car. The Corvair, also an air cooled, rear engine car, but with six cylinders instead of four to outperform its German rival. Corvairs were built from 1960-1969 with a major body change in 1965.

So my search was redirected to Corvairs and in Southern California they were getting top dollar. I found a 1966 four door Corvair on Ebay in Modesto California listed as 'not running' which was music to my ears. The body, faded but damage free, was also very enticing, but a four door muscle car?

I bid $610.00 and won it, then rented a car trailer and made the six hour drive north. I didn't realize until I picked the Corvair up that the roof was dramatically different between the two and four door versions. I liked the more Camaro look of mine, the four door.

Two Door Roof   Four Door Roof

WHERE IT BEGAN

So you're not seeing my vision of this beauty becoming a unique muscle car? I took this side view photo to help me visualize on the computer a 'What if" shortened version and produced the next photo as an extreme possibility.

A little radical, but it fired me up having a respectable two door version of a muscle car. Another thing I'd like to point out, it was very popular in the 70's to purchase a Crown Conversion kit that would allow you to put a V8 engine in the back seat of a Corvair. I was putting my V8 up front which allowed me to consider shortening the body striving for something different. Out with the old...

Front suspension   **Corvair rear 6 cyl. engine**

The magic dimension for how much to shorten the body came from the rear door. The side body lines went 15 inches along the back door before transitioning up so 14" would make matching up much easier.

Knowing I was going to shorten the body I needed to fabricate a crude fixture that would allow me to cut through the body and slide the front and back sections together for welding. The 4X4's provided a perfect guide.

A saws-all with a fine metal cutting blade did the job perfectly through the back seat area and the roof section up front for a better contour match. At this point, I didn't know if I would weld the rear doors into the body or leave them usable.

To aid in alignment, add strength and backup the welding, I made inserts that would slide into the body cross sections as shown below. Note the high tech ratchet straps used to bring the two halves back together.

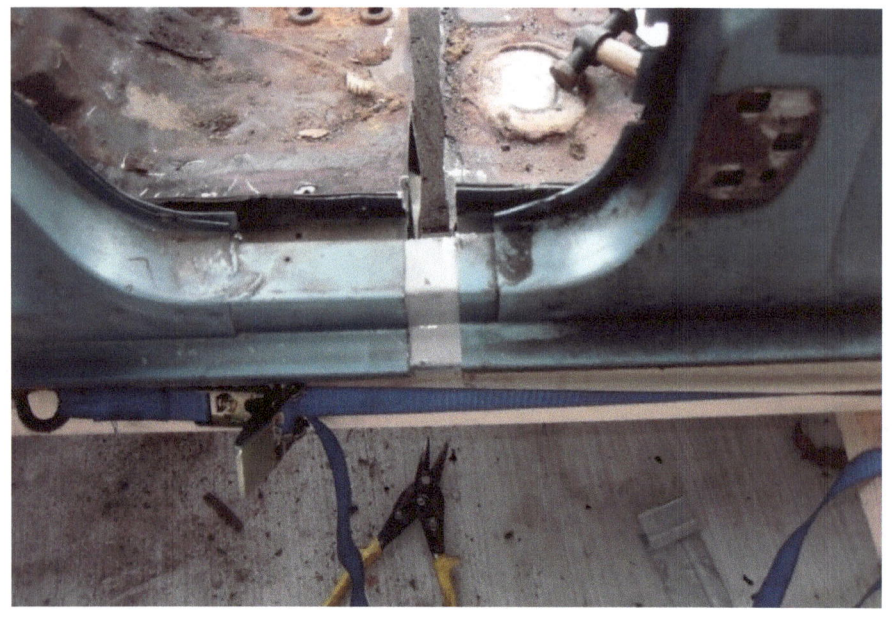

Inside the roof, a 1/8" thick X 3/4" wide flat steel strip was used to both backup the weld and resist buckling.

Aligned, clamped and fourteen inches shorter, it's ready for MIG welding back together. Shown below, the body was welded and lifted in preparation for the next step of adding a chassis.
(What is MIG welding? See page 33)

The Corvair is a uni-body construction which means it didn't have a separate body and chassis. I knew the front suspension was designed to support quite a few bags of groceries up front, but would probably not be well suited for the coming modified Chevy 350 V8. So it was time to find a chassis to put under my Corvair. I could easily make the chassis longer or shorter and focused on the required width. My research gave me the Chevy S-10 series would be the perfect width and searching for chassis' was pointless as many of the other components that I'd need would have to be purchased separately, driving the cost through the roof. The good Lord directed me to one of those donate your car auction sites and gold was found.

Chapter 2

It was an S-10 Chevy Blazer that someone had the thought a fresh coat of paint would reset the two hundred thousand mile odometer reading. Again listed as 'not running', it was a steal at $400 and that new paint job got my cousin, Steve excited to help me dismantle the body for fenders, grill, bumper, hood and doors for his truck.

Speaking of all of those extra components, what was exciting was my Corvair would have power front disc brakes and power steering. Not to mention this Chevy Blazer was providing the rear differential and drive shaft I'd need.

Days later the Blazer was stripped and we pulled the chassis free.

I probably don't need to point out that this was a home project; my Uncle Paul's auto shop was long gone and improvise was the theme song for this project. Once the chassis was free, it needed to be shortened 8-1/2 inches for my Corvair wheelbase and I was confronted with another adjustment needing to be made. The Corvair was designed for 13" wheels and tires and the Blazer chassis has 15". Cutting out the wheel wells on the Corvair was not an option to keep it's appearance in the sleeper category. I purchased the lowest profile 15" tires I could find and was ready for bringing the body and chassis together. Love those 4X4's.

Using the engine hoist and taking several steps front and back the body was lowered onto the chassis. In preparation, I'd removed the floor board in the Corvair body to allow the chassis inside. The chassis was leveled and the body supported with electric screw jacks to also level it and the two were welded together at contact points and brackets made to bridge gaps in key locations.

I had the same thought, good thing I ran out and got those low profile tires to fit in these wheel wells! NOT!

No problem at all, setting that heavy 350 V8 engine and transmission in place will drop her like a rock.

That's Chris, by the way. It was early in the project and she was both helpful and still talking to me at that time. This was the engine and transmission set in place before rebuilding to create the motor and transmission mounts and the form I needed to build around.

Yeah, right. Maybe not drop it like a really big rock, more like a little stone. Who said get a truck chassis? Me, Oh well…

With the engine temporally in place, let's take a look at how those must have muscle car headers fit.

As you will see one the next page, the header and frame both required modification for clearance.

I could have sworn that the header catalog indicated a perfect fit for a Chevy 350 in the front trunk of Corvair body mounted on a Blazer chassis? Oh well, again …

Help needed on the driver's side …

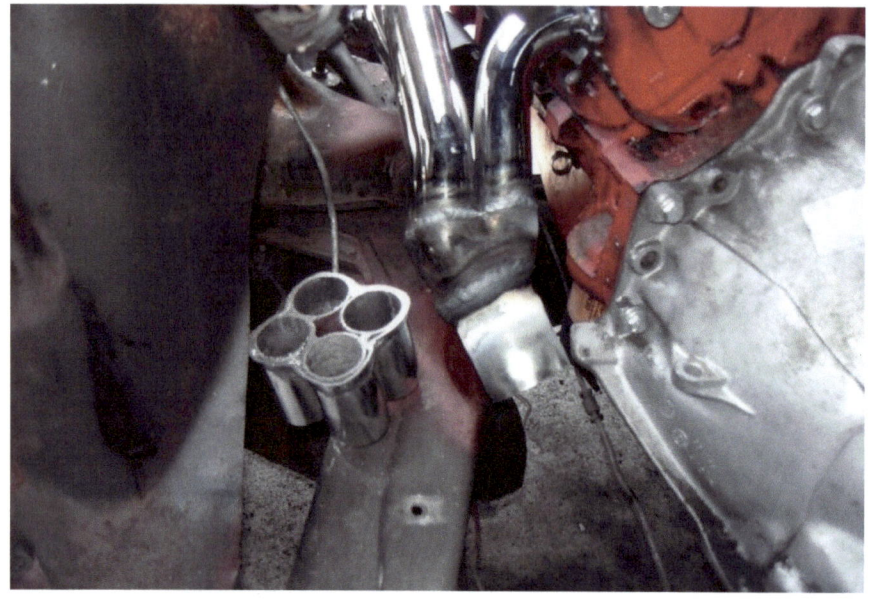

… slightly more the passenger side.

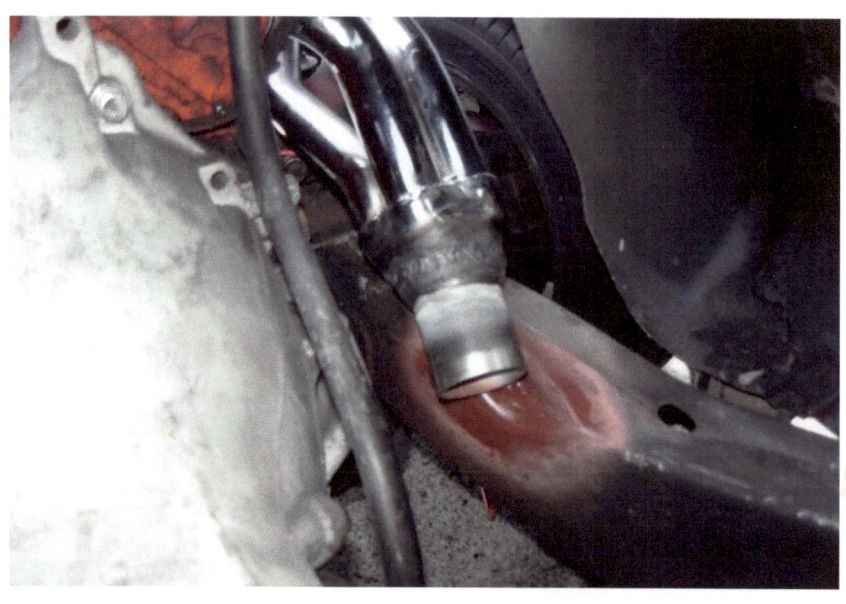

Realizing the weight of the engine and transmission wasn't going to bring my Corvair back to earth, I used a combination of 2" drop spindles and 1" drop springs in the front and 3" drop blocks in the back. That was also a good time to replace all of the front-end swing arm bushings as well. With the 3" lowering blocks in the back on leaf springs and anticipating more torque would find it's way back there, I added traction bars during the process.

Looking inside the car and seeing the driveway, maybe it was time for a floor board. I glanced into the backyard where the carcass of Chevy Blazer still sat and smiled.

Saws-all in hand, I went and cut the floor board out of the Blazer in two sections, that made fitting it into the Corvair remarkably easier to trim and weld into place.

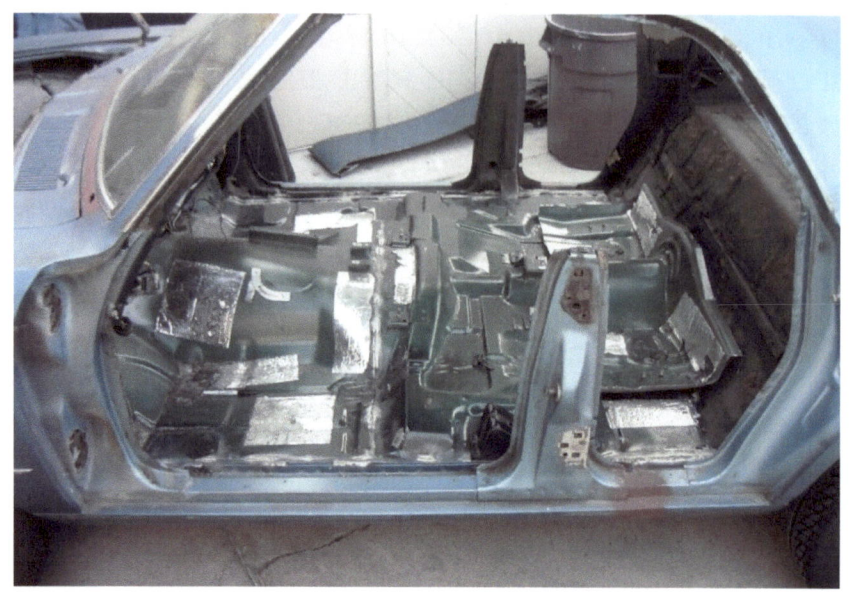

Happy dance time! My floor was easily installed, fit like a glove and welding was a breeze.

Then I put the Blazer seat into it and for the first time in my life wished I was five foot seven, not six foot three. I put the old, lower Corvair seats in and only needed to be two inches shorter. LESSON, tack weld things first, check it all out thoroughly and then go crazy welding all the seams...

*****

Floor board #2, AKA, the art of cutting patterns and building puzzles with the only benefit being,

I would be able to drive my Corvair!

I was able to still use the front section of the Blazer floor board to wrap around the transmission and built the remaining floor support using channel where the seats would be bolted down and a half section of 6 inch steel pipe for the drive shaft tunnel.

Once the floor support frame was in place it was time to cut cardboard patterns, trim and fit the flat sheet steel sections to be welded into a one piece floor board.

The firewall sections and funnel shaped transition for the driveline are shown going into place above.

The taller Blazer seats were still out, but honestly, I really wanted an excuse to upgrade to better seats anyway and found these high back seats on sale! We have a floor board!

Chapter 3

Pulling the engine back out of the car, it was time to start the rebuilding process for the muscle part of this Corvair project.

It was like opening up one of those mystery packages when you're tearing an engine down to see what's inside.

Surprise! My first prospective engine apparently held some water ...

What do they say, if at first you don't succeed, try, try, try and then try again? Yeah, it took 5 tries to get to a 350 block I could rebuild. Number two was already bored .040 over and worn out, three had a broken piston and sleeve and number four was a block the engine shop had offered, but after boring the cylinders he discovered the main caps didn't line up and wanted an additional $250.00 in machining expense. I declined his offer and finally found in my fifth attempt a rebuild able engine block to send to the machine shop.

Once cleaned, bored .040 over, surfaced and the cam bearings installed, I was ready to go.

Running a tap through all of the threaded holes to clean them out is always a good idea.

For those of you that haven't looked this deeply into an engine, I'll add a few assembly photos.

First, setting the reground crankshaft into the block.

New 9:1 compressions pistons.

A moderate camshaft was chosen for lift and duration just below needing a racing torque converter and of course, stronger valve springs and new lifters.

An Edelbrock performer manifold and carburetor were chosen along with Edelbrock dress up stuff, valve covers and air filter.

Edelbrock was my first drafting job, two-and-a half years. Just sayin' ....

Before I could put the completed engine back into its new home, no longer the trunk of my Corvair, there was some up front work to be done.

In the rough, here are the mated and welded sections of the firewall and closing the wheel wells around the upper swing arms.

Looking forward now in the engine compartment there are the welded side panels for the coming radiator.

At the top of this photo, you can see the trunk latch,

a bonus now, by putting the engine in the trunk, I had a key locking engine compartment!

First pass with primer and paint up front. Yeah, maybe I was in a hurry at this point and did spend more time dressing this area up, making the welds disappear about after a year or so. Upper right the wheel well was modified for the power brake canister. The mating brake pedal assembly was another component taken out of the Blazer and fitted in the Corvair.

I couldn't use the engine block, ol' rusty from that initial purchase, but the 350 three speed automatic transmission that came with it was used to get me moving in the beginning. I purchased a rebuilt 700R-4, that offered overdrive a couple of years later. Quite a difference on the freeway.

The engine and power brake master cylinder are in place and the bottom radiator air scoop.

Looking forward again is the completion of radiator enclosure.

Not shown here are the two 12" electric fans that

were attached to pull air through the radiator. It was too low in relationship to the engine for attaching the fan commonly bolted to the engine's water pump. Unlike rolling the dice that the purchased used transmission would work, I bought a new aluminum dual core radiator instead of fitting the original Blazer one in.

Moment of truth, short of chopping 14" out of the body, it was my goal to conceal the obvious fact that this Corvair had a V8 engine up front. I had already decided I was going to leave the gas tank access door in the front fender as it aided in the illusion. But the real key to achieving a 'sleeper' was bringing the trunk lid, now the hood over to see if it would close without the need of a bump in the middle for air cleaner clearance.

(sleeper = Shhhhhh it's a V8)

Short, only to the thrill of hearing the lope when I first started the engine, bringing that hood down without interference was my greatest thrill. You'd think being able to sit in it would have been a topper, but quite frankly, I'd rather sit on the floor board than destroy the stock lines of the hood.

Road tests proved the transmission was sound, thank you, Lord, but also that the bottom air scoop alone was not letting enough air through. The most efficient answer would have been to simply open up the front, but again being a 'sleeper' was important to me so the first step was to work below the front bumper.

Cutting in and welding two tunnels in below the bumper proved to be enough additional air flow.

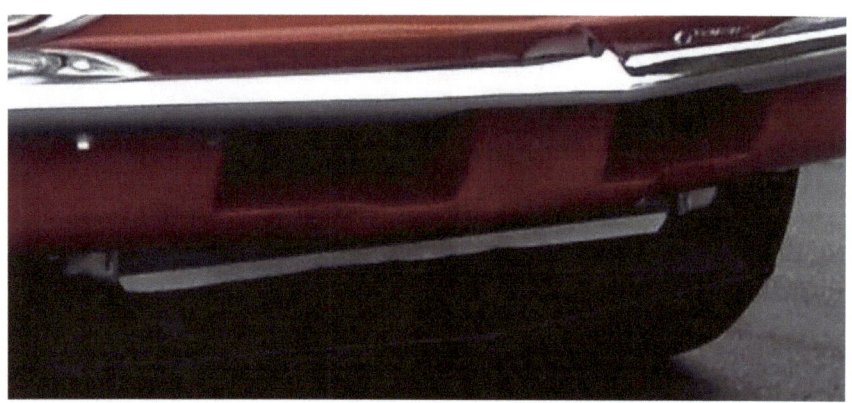

The front doors had been installed without any need of modification so the 14" removed from the body was all taken out in the rear doors. Having the door handles so close together was okay for test driving, but looked a little silly. Two choices; both removed the rear door handle, but either I could weld and blend the rear door into the body or put electric solenoids in the rear doors and keep them.

I chose the latter and it surprised me at car shows later that those short functioning rear doors brought as much comment as a Chevy 350 in the trunk of my Corvair.

Another note here, as the window glass was tempered, I couldn't have it cut down to the new required size. I had to cut rear door windows out of 1/4" Plexiglas.

As every backyard mechanic will attest to, any project that requires the purchase of new tools is a good project. Twist my arm, yeah right, but welding the trimmed door skins back to the shortened door frames required a TIG welder.

You pros can jump ahead here to the next chapter. For those of you not knowing the difference between welding and soldering here is an overview of welding 101.

WELDING 101

First, soldering for joining metal parts is heating the metal and applying lead solder, like hot glue. Usually an electric iron or propane torch is used to heat the metal. A grease like flux is used to assist and the key is to heat the metal hot enough to melt the solder on contact, not drop melted solder on the parts. All copper plumbing is joined this way.

Welding is actually melting the metal to join two parts together.

- A. Oxy-Acetylene welding is using a torch handle that blends oxygen and acetylene from different cylinders together to support a flame that melts metal. A metal rod is used to dip into the puddles of molten metal to create a joint bead between the parts. The rod is used up to fill the joint.

- B. Arc welding uses electrical voltage not a gas fed torch. The ground lead is clamped to the parts being welded. The handle holds an electrode, a metal rod coated in a flux material about a foot long. When the tip of the electrode contacts the parts grounded an electric arc is created, melting the metal parts and consuming the electrode as filler.

- C. MIG (Metal Inert Gas) welding takes arc welding to the next level. First, instead of continually replacing electrodes as they melt into the parts being welded, a MIG welder has a spool of wire that is feed through the handle. The tip of the wire creates the electric arc. A shielding gas also flows through the end of the handle where the wire is feed out to push oxygen and water vapor away from the welding area.

D. Wire welding is done with the same MIG welder as in C above, but does not use a shielding gas. Inside of the fed wire is a flux that during the welding process creates a gas keeping oxygen away from the joint bead during welding.

E. TIG (Tungsten Inert Gas) welding is used for thin sheet metal welding. Again an electrical process where the parts being welded are grounded and a shielding gas is used. The handle for TIG welding has a small diameter tungsten tip to create the electrical arc that melts the metal and a separate hand held rod is fed into the weld bead. This process keeps the arc focused in a small area to avoid distortion in the metal beyond the welded area.

Chapter 4

Mechanically, all was right with the world and I took the body prep as far as I could go to get it ready for a face lift at the paint shop.

The paint shop has a process in their deluxe paint package where they paint the car and you and the manager, armed with a grease pencil, points out the imperfections and circles the ones you want repaired. I had already decided on an orange-red color I'd seen on the new Camaro called "Inferno Orange", but this gave me a chance to spend more money on fine tuning the body and to see my car in deep blue paint.

A few days later I got a call that the color I wanted required a special process that they were unable to do. I went down to go through their paint books and found an old Chrysler color that was close. Back home with the chrome and trim back on, it was time to do something about those wheels and finish up the inside of the car.

I was thrilled with these wheels, different and tied into the shape and style of the narrow, rounded bumpers.

And inside, I used a Saturn center console.

Those 'on sale' high back bucket seats.

Those fun rear door and a sun roof!

Finished under the hood.

The old engine compartment, now trunk.

My first car show

'California Jam at the Park' in June of 2010 was my first car show and we won two trophies, 'Top 50' and one in a more specific class, 'Best Engineered'.

Had to take the story with me.

That show was the first of many car shows and I thoroughly enjoyed every step of the process of building something unique. The fun times were at stop signs and the double takes, many I'm sure, had no idea what a Corvair even was.

Letting my Corvair go three years ago was like losing a child and even now when I drive by some odd piece of automotive history in someone's back or side yard I think, I could do it again.

And with that, it's time to leave you.

It's now 2017 as I put these memories on paper, hopefully some of you will be inspired to leap into a project and I wanted to dedicate it to my six grandchildren, a look back at their grandpa.

Thanks for reading this and I hope you enjoyed it. On the next page are some of my fiction works available on Amazon or Barnes and Noble in Ebook or paperback. Yeah, I'm still welding.

Watson Manor Mystery Series

Watson Manor Eventually
Watson Manor Unfolding
Watson Manor Investigations
Watson Manor My Journey Home

Novellas

In His Time
Katie dear's Resolve

Written for a Student

Kye's First Car

www.ingramcontent.com/pod-product-compliance
Lightning Source LLC
Chambersburg PA
CBHW041113180526
45172CB00001B/229